HOW TO EXCEL IN DATA SCIENCE INTERVIEW

Re-Occurring Interview Questions And Answers To Make You Get Good Grades And Champ The Quiz, 2018 Updated

By

KEN J. STEVENS

TABLE OF CONTENTS

Ken J. Stevens

CHAPTER 1

A QUICK DIVE TO UNDERSTAND THE CONCEPT OF DATA SCIENCE

ANOTHER AREA SURFACES there may be huge and growing interest for data astute experts in agencies, open places of work, and charities. The delivery of professionals who can work adequately with data at balance is constrained, and is pondered by fast growing pay costs for data engineers, scientists, and analysts.

AN AWESOME EXPLOSION OF DATA

data is increasingly more cheap and ubiquitous. We're now digitizing analog content material that turned into created over masses of decades and accumulating myriad new kinds of records from net logs, cellular gadgets, sensors, devices, and deals. IBM estimates that ninety two percentage of the records in the world these days has been created inside the past couple of years.

Within the same time, new technology are emerging to prepare and appear practicalof this avalanche of data. We should now pick out conduct and regularities in data coming from all kinds that permit us to enhance scholarship or furnish, growth the human circumstance, and create industrial and interpersonal price. The upward push of "big statistics" has the potential to deepen our expertise of phenomena varying from physical and natural systems to human interpersonal and financial behaviour.

A TROUBLE RECOGNIZED

in reality every zone of the financial system now has get right of entry to more data than would were imaginable even a decade in the past. Groups today are gathering new records at a charge that exceeds their capability to extract cost from it. Hassle going through each company that desires to capture the eye of a community is the way to use dataefficaciously -- now not honestly their own information, but all of the data that may be discovered and relevant.

Our ability to uncover social and financial fee from the newly available data is restricted with the aid of the dearth of experience. Working collectively with this information calls for special new abilities and tools.

The corpuses are regularly too voluminous to suit on a sole laptop, to control with conventional databases or record tools, or to signify the use of preferred snap shots software. The facts is also greater heterogeneous than the rather curated records of earlier times . Digitized textual content, sound , and visible content material, like sensor and blog information, is normally messy, incomplete, and unstructured; it's miles regularly of unsure provenance and fine; and regularly must be blended with different records to be beneficial. Working collectively with consumer-generated information programs additionally increases difficult problems of privacy, safety, and ethics.

The sphere of statistics science is developing on the intersection of the fields of interpersonal science and facts, facts and pc science, and layout. The UC Berkeley school data is preferably situated to deliver these physical games together additionally to offer college students with the research and expert competencies to do properly in main side businesses.

CHAPTER 2

GETTING PREPARED FOR THE INTERVIEW

Setting up for a task interview isn't clean - evidently there can be a massive quantity of uncertainty about the data technology interview questions you will be requested. No matter how lots paintings revel in or technical talent you have got, an interviewer can positioned you off with a fixed of questions which you failed to count on.

Referring to a statistics technological know-how interview, an interviewer will inquire approximately spanning a ramification of subjects, necessitating sturdy technical know-how and communique skills from hassle the interviewee. Your reports, programming, and statistics constructing talents could be put to the take a look at through a diffusion of questions and query styles - intentionally designed to maintain in your toes and pressure you to show how you will operate below strain. Guidance is a major key to success whilst on the lookout for a profession in data technology.

This guide contains all the statistics science interview questions an interviewee have to count on when interviewing for a role as a statistics scientist. At Springboard, we train facts technological know-how through our mentored records medical studies workshops. They're a fantastic way the first-class manner to research data technology and get professional guidance on getting a statistics science job. We did our predicted diligence to comb via the net to locate actual questions asked to facts technological know-how interview prospects. We had built a statistics science interview guide, yet we still sensed we had more to discover.

We activate to curate, create and edit distinctive information scientific studies interview questions and provided answers for some. Using this set of statistics technological know-how interview questions, an interviewee could be able to prepare for the challenging questions, examine what solutions will positively resonate with an company, and expand the confidence to ace 250 the interview. We've cracked the facts technological know-how interview questions into six different categories: facts, programming, constructing, conduct, lifestyle, and hassle-solving.

CHAPTER 3

STATISTICS

Statistical work is the manner via which statistics scientists take raw statistics and create predictions and models reinforced by way of the data.

Without a sophisticated expertise of records it's miles tough to acquire fulfillment as a data scientist - for this reason it's miles probably an awesome interviewer will try to probe your information of the problem depend with information-oriented facts technology interview questions. Be prepared to reply some essential reviews questions as a part of your facts technology interview.

Right here are examples of rudimentary facts questions we've observed:

precisely what's the vital restrict Theorem and why is it essential?
What's sampling? What number of sample techniques do you know?

Exactly what's the distinction among type i vs type II mistake?

What's linear regression? So what do the situations P-value, coefficient, R-Squared value mean? Precisely what is the importance of every of those components?

What are the presumptions important for linear regression?

What is a report interplay?

What is selection bias?

Exactly what is an instance of a dataset with a non-Gaussian distribution?

Precisely what is the Binomial chance formula?

CHAPTER 4

PROGRAMMING

To check your coding competencies, employers will ask 2 things all through their information technology interview questions: they'll ask how you will possibly remedy programming troubles in concept without writing out the code, after which they may additionally provide whiteboarding sporting events so you can programme or code right now.

For the latter forms of questions we can cover a few true examples under, whilst you're searching out in-intensity exercise coping with coding challenges, visit Interview Cake. They have got an in-browser module for keying in code, and they can stroll you through complicated troubles - all certainly unfastened.

CHAPTER 5

GENERAL

With which programming languages and conditions are you most comfortable working?

Precisely what are some advantages and cons about your selected statistical software program?

Inform me for my part about an unique formulation you've got created.

Describe a facts technology mission in which you labored with a huge programming factor. What did you discover from that experience?

Perform any begin supply tasks?

How would you smooth a dataset in (insert language here)?

Inform me approximately the coding you did all through your ultimate venture?

CHAPTER 6

LARGE DATA

Exactly what are the 2 primary components of the Hadoop Framework? Give an explanation for how MapReduce works as in reality as it could be.

How could you type a big set of numbers? The following is a huge dataset. What is your coverage for handling outliers? What about missing values? How about adjustments?

CHAPTER 7

PYTHON

What modules/libraries are you maximum acquainted with? So what do you want or detest facts?

What are the supported information kinds in Python?

Exactly what is the difference among a tuple and a listing in Python?

For additional Python questions that provide attention to looking at precise mind of code, take a look at away this useful aid created by way of Toptal.

CHAPTER 8

R

What are unique varieties of sorting algorithms available in R language?

There are insertion, bubble, and selection sorting algorithms.

Precisely what are unique facts objects in Ur3rd there's r?

What packages are you maximum acquainted with? And so what do you want or dislike facts?

Simply how do you get right of entry to the aspect inside the 2nd column and 4th line of a matrix known as M?

What is the command used to store R gadgets in a document?

Precisely what's the satisfactory manner to use Hadoop and 3rd there's r collectively for evaluation?

Just how do you cut up a continuous variable into exceptional businesses/ranks in R?

Compose a characteristic in Lthird there's r language to update the lacking value in a vector with the thought of that vector.

CHAPTER 9

SQL

Often, sq. Questions are case-based, which means that an business enterprise will assignment you with solving an sq. Hassle so as to test out your capabilities from a practical point of view.

For example, you could be given a stand and be asked to extract relevant records, filtration system and order the information as you spot in shape, and document your effects. In case you do not sense ready to do that in an interview placing, Mode Analytics has a delightful introduction to the usage of sq. In an effort to educate you these instructions via an interactive sq. Environment.

What's the reason of the organization features in sq.? Offer some examples of institution capabilities.

Group features are important to get short summary data of your dataset. Count number wide variety, MAX, MIN, AVG, amount, and wonderful are all group functions show me individually the distinction among an inner be a part of, left be part of/right join, and union.

What does UNION do? What is the difference among UNION and UNION nearly ALL
what is the big difference between sq. And MySQL or square Server?
If possibly a table consists of equal rows, does a question result display the reflect values by means of default? Just how can you eliminate mirror rows from a question result?

CHAPTER 10

TROUBLE-SOLVING

Interviewers will, in some unspecified time in the future all through interview manner, need to test your problem-solving ability through information medical research interview questions. Frequently these checks can be furnished as an open-ended query - How might you do "X"? Normally talking, that 'X' might be a task or problem precise to the agency you are making use of with. Intended for example, an interviewer at Yelp might also ask a prospect how they might create a device to find out faux Yelp critiques.

Some short hints: avoid be afraid to invite questions. Employers need to test your important questioning competencies - and requesting questions that clarify parts of uncertainty are a top notch manner to expose you recognize the way to ask the proper questions (a trait that any facts scientist need to have). As nicely, if the mission offers an opportunity to show off your white-board coding talents or create schematic diagrams - use that for your benefit. It indicates technological skill, helping to attach your way of thinking about via a unique medium of conversation. Always talk your concept system -- process is frequently extra critical than the outcomes themselves for the task interviewer.

How would you come up with an answer approach to perceive plagiarism?

How many "beneficial" votes will a Yelp evaluation receive?

How are you going to discover person paid accounts disbursed by means of a couple of customers?

You're about to ship 1,000,000 emails. Just how do you optimize transport?

How can you optimize response?

You have a dataset that contains100K rows and 90 columns, with one of these columns being our dependent variable for troubles we'd like to resolve. How can we fast perceive which columns can be beneficial in forecasting the based variable. Determine techniques and describe them to me as although I had been 5 years old.

How would you discover bogus opinions, or bogus facebook facts files used for bad functions?

This is an opportunity to show off your know-how of machine gaining knowledge of algorithms; particularly, sentiment evaluation and textual content evaluation algorithms. Showcase your understanding of fraudulent patterns - what are the ordinary behaviors that could commonly be visible from fraudulent debts?

How could you carry out clustering on 1,000,000 specific keywords, assuming you've got 12 million records points - every one along with key phrases, and a metric measuring how comparable those keywords are? How would you create this 10 million facts factors table within the first location?

How could you optimize an internet crawler to run plenty faster, extract higher records, and better summarize facts to produce cleaner information supply?

CHAPTER 11

WAY OF LIFE

If an employer asks you a question in this list, they're looking to get a feel of who you're and how you will supplement the organisation. They're trying to gauge in which your hobby in statistics science and the using organization come from. Check these examples and think about what your first-rate answer might be, but maintain in mind - it is essential to be honest with these questions. There may be no cause to no longer be yourself.

Currently there aren't any right solutions to these questions - but the nice solutions are communicated with self-confidence and a smile.

Which statistics scientists do you respect most? Which on-line corporations?

What do you suspect makes a very good information scientist?

How did you come to be enthusiastic about facts science?

Supply some examples of "best practices" in records technological know-how.

What/when is the ultra-modern data science e-book / article you read?

What/while is the modern facts mining convention web seminar/ class / workshop / training you attended?

What challenge you will need to paintings on at our agency?

What particular abilities do you believe you studied you'd deliver to the crew?

What statistics might you want to acquire if there were no constraints?

Have you ever ever notion approximately developing startup? About which principle?

What can your interests and interests inform me that your resume can't?

What are your top 5pinnacle 5 predictions for the next two decades?

What did you do these days? Or what did you do that week in addition to remaining week?

In case you gained a million greenbacks in the lottery, what would you do with the money?

What is one aspect you believe you studied that the general public do now not?

What persona traits do you truly butt heads with?

What are you enthusiastic about?

CHAPTER 12

SUMMARY

There may be no precise method for getting ready for statistics technology interview questions, however hopefully by way of searching at these common interview questions you will be able to stroll into your interviews well-practiced and confident.

THE END